Too Cute!
Baby Raccoons
by Elizabeth Neuenfeldt
BLASTOFF! Beginners
BELLWETHER MEDIA
MINNEAPOLIS, MN

Blastoff! Beginners are developed by literacy experts and educators to meet the needs of early readers. These engaging informational texts support young children as they begin reading about their world. Through simple language and high frequency words paired with crisp, colorful photos, Blastoff! Beginners launch young readers into the universe of independent reading.

Sight Words in This Book

are	how	on	this
at	in	one	to
find	is	the	up
get	it	their	
have	look	them	
help	many	they	

This edition first published in 2023 by Bellwether Media, Inc.

Library of Congress Cataloging-in-Publication Data

Names: Neuenfeldt, Elizabeth, author.
Title: Baby raccoons / by Elizabeth Neuenfeldt.
Description: Minneapolis, MN : Bellwether Media, 2023. | Series: Blastoff! beginners: Too cute! | Includes bibliographical references and index. | Audience: Ages 4-7 | Audience: Grades K-1
Identifiers: LCCN 2022012984 (print) | LCCN 2022012985 (ebook) | ISBN 9781644876718 (library binding) | ISBN 9781648347177 (ebook)
Subjects: LCSH: Raccoon--Infancy--Juvenile literature.
Classification: LCC QL737.C26 N48 2023 (print) | LCC QL737.C26 (ebook) | DDC 599.76/32--dc23eng/20220325
LC record available at https://lccn.loc.gov/2022012984
LC ebook record available at https://lccn.loc.gov/2022012985

Editor: Christina Leaf Designer: Jeffrey Kollock

Printed in the United States of America, North Mankato, MN.

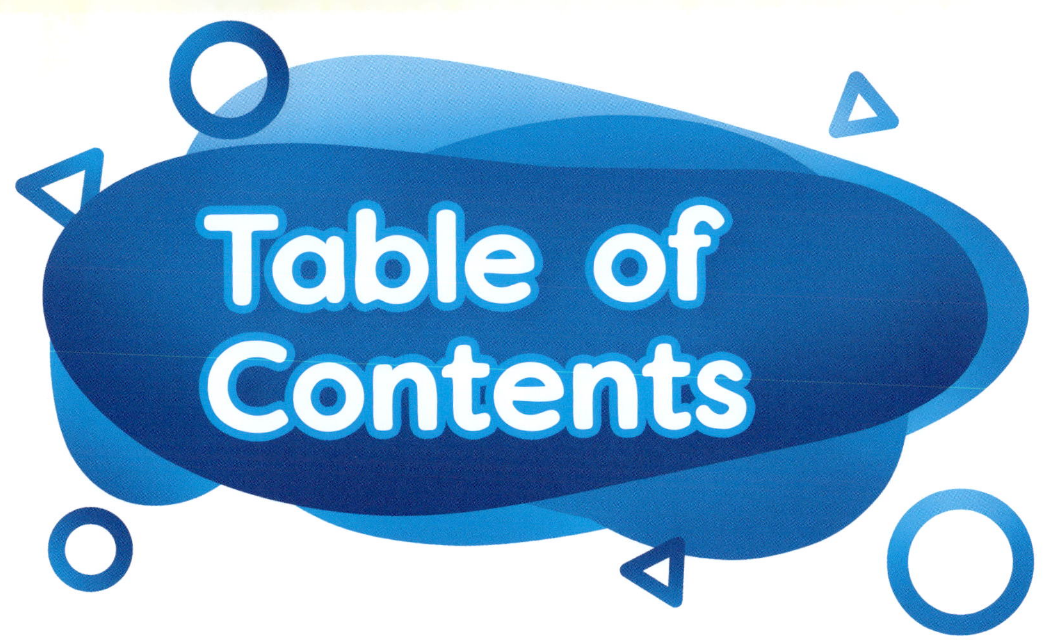

Table of Contents

Look at the baby raccoon. Hello, cub!

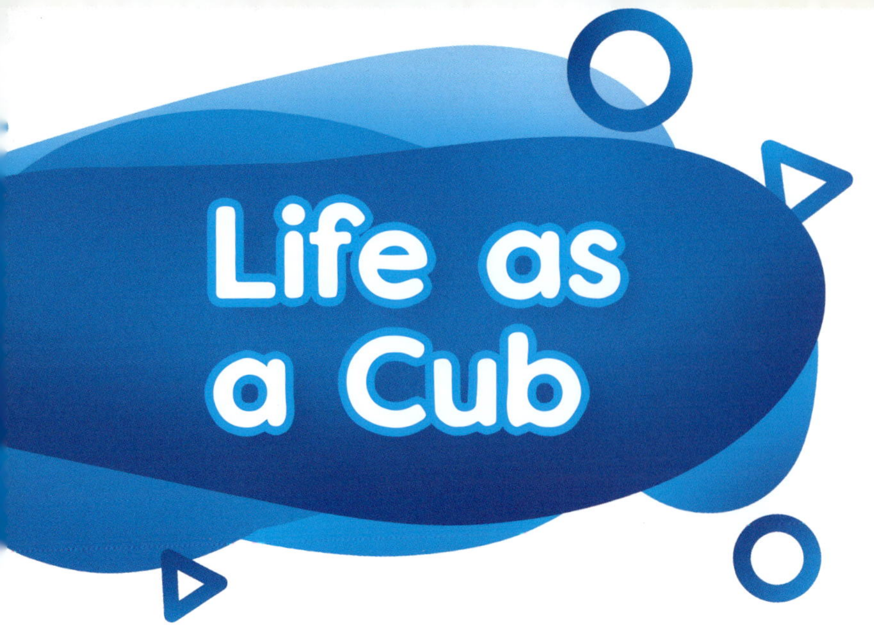

Life as a Cub

Newborn cubs have dark fur.

Cubs live
in **dens**.
Dens are in
logs or trees.

den

Cubs have many **siblings**. They snuggle to stay warm.

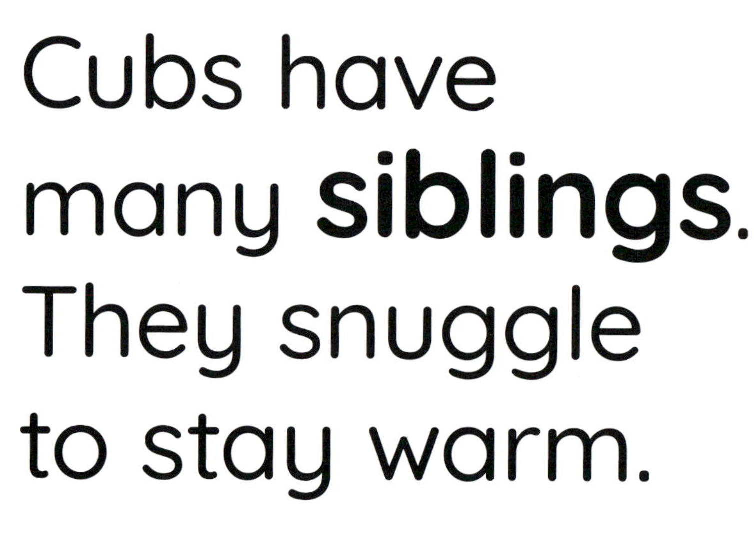

siblings

Cubs drink
their mom's milk.

drinking milk

Growing Up

Older cubs learn how to find food. Mom shows them.

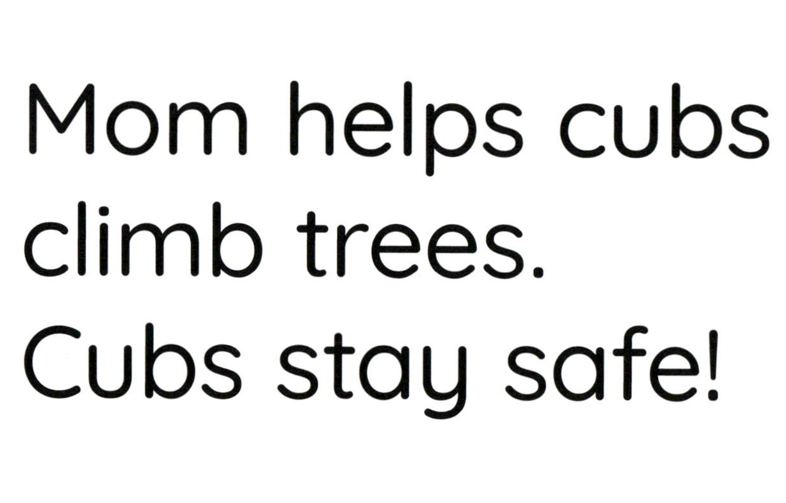

Mom helps cubs climb trees. Cubs stay safe!

Cubs get bigger. They **explore** on their own.

This cub is one.
It is grown up.
Goodbye, mom!

Baby Raccoon Facts

Raccoon Life Stages

newborn cub adult

A Day in the Life

learn to find food

climb trees

explore

Glossary

homes for raccoons

to look around a new place

just born

brothers and sisters

To Learn More

ON THE WEB

FACTSURFER

Factsurfer.com gives you a safe, fun way to find more information.

1. Go to www.factsurfer.com.
2. Enter "baby raccoons" into the search box and click 🔍.
3. Select your book cover to see a list of related content.

Index

The images in this book are reproduced through the courtesy of: Sonsedska Yuliia, front cover, pp. 5, 22 (cub); Eric Isselee, pp. 3, 4, 10, 22 (newborn, adult); Design Pics/ SuperStock, pp. 6, 23 (newborn); Holly Kuchera, pp. 6-7; Figtography, pp. 8-9; Betty Shelton, pp. 10-11; C.C. Lockwood/ agefotostock, pp. 12-13; Duncan Usher/ Alamy, pp. 14-15; outdoorsman, pp. 16-17; Debbie Steinhausser, pp. 18, 23 (explore); Don Johnston/ Alamy, pp. 18-19; Lubomir Novak, pp. 20-21; Evelyn Harrison/ Alamy, p. 22 (food); Heiko Kiera, p. 22 (climb); Agnieszka Bacal, p. 22 (explore); Vladimir Turkenich, p. 23 (dens); Geoffry Kuchera, p. 23 (siblings).